Learning MULTIPLICATION with Puppies and Kittens

Linda R. Baker

Enslow Publishing
101 W. 23rd Street
Suite 240
New York, NY 10011
USA
enslow.com

Words to Know

basic Simplest.

equal Of the same value or size.

factor One of the numbers that are multiplied.

multiply A fast way to add two or more objects of the same amount together.

operation A way to get one number using other numbers by following special rules.

product The answer to a multiplication problem.

value The amount or worth of something, such as a number.

Contents

Multiplication

Multiplication is a kind of math you use to add the same number over and over again. It is a fast way to add. The two numbers that you **multiply** are called **factors**. The **product** tells you how many items you have altogether. You can count puppies and kittens to learn how multiplication works!

2 × **4** is a multiplication fact.

Multiplication facts can be written in several different ways:

$$2 * 4 = 8$$

$$2 \bullet 4 = 8$$

$$2(4) = 8$$

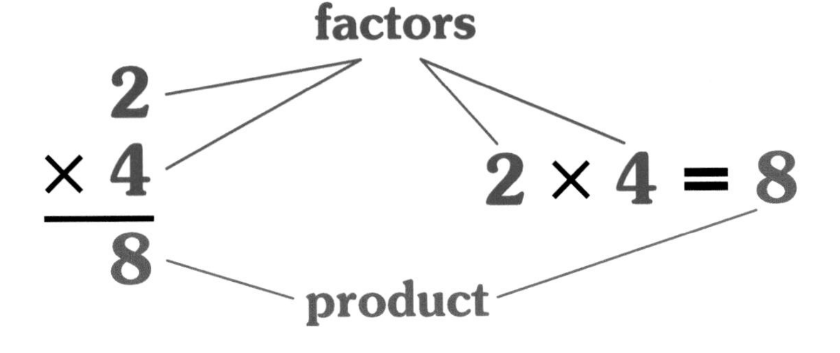

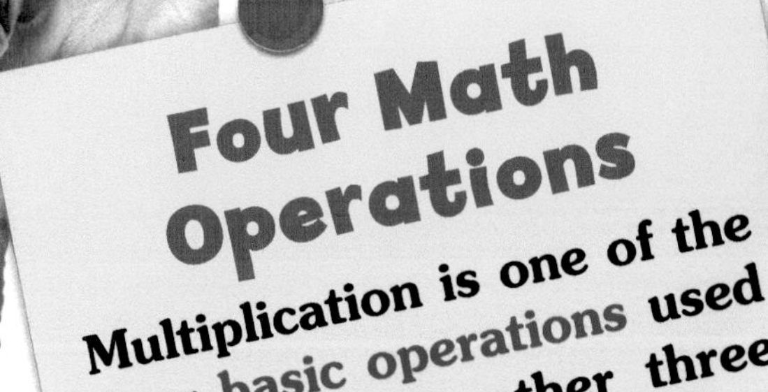

Four Math Operations

Multiplication is one of the four basic operations used in math. The other three are addition, subtraction and division.

There are four parts to a multiplication fact:

(1) The two (or more) factors are multiplied together: **2** and **4**.

(2) The multiplication sign (×) means to multiply.

(3) The **equal** sign (= or __) means solve the problem.

(4) The product, or answer, comes after the equal sign: **8**.

Learn to Multiply by 0

When you multiply any number by **0**, the product is always **0**. Zero has no **value**.

There are **0** puppies in any of the **10** puppy cages at the shelter. How many puppies are in all the cages at the shelter?

All the puppies have found homes, so there are **0** puppies in the cages!

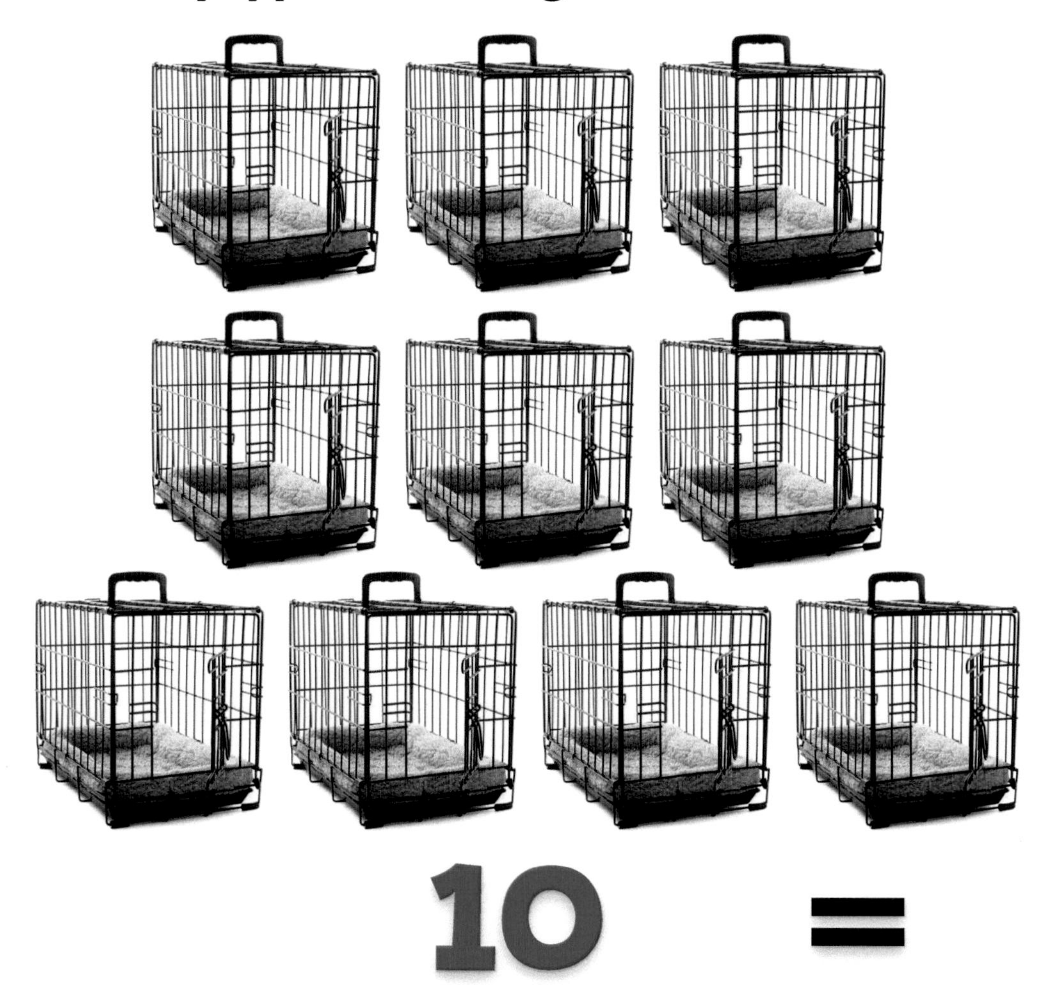

$0 \times 1 = 0$
$0 \times 2 = 0$
$0 \times 3 = 0$
$0 \times 4 = 0$
$0 \times 5 = 0$
$0 \times 6 = 0$
$0 \times 7 = 0$
$0 \times 8 = 0$
$0 \times 9 = 0$
$\mathbf{0 \times 10 = 0}$

10 = 0

Learn to Multiply by 1

When you multiply any number by **1**, the product is always that number.

There is **1** big cat tree for **6** kittens to climb. How many kittens are climbing the cat tree?

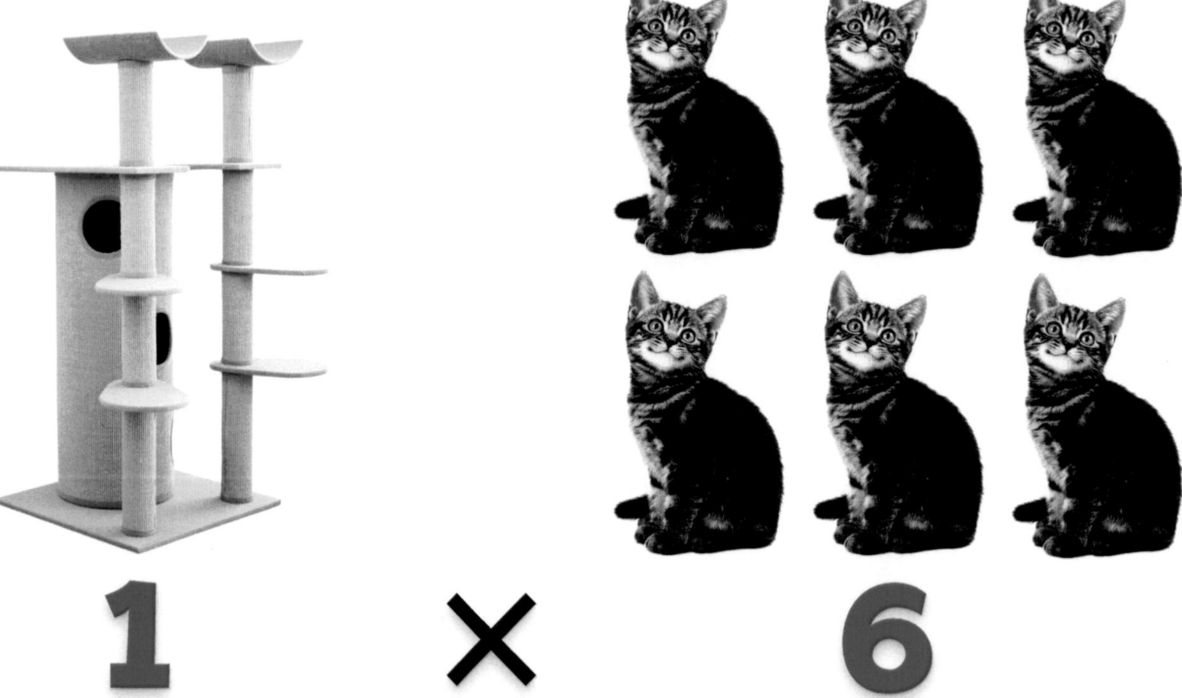

Commutative Property

Unlike division or subtraction, the order of the factors in multiplication can be changed and the product will be the same: $1 \times 6 = 6$ and $6 \times 1 = 6$. This is called the commutative property of multiplication.

There are **6** kittens climbing the cat tree.

1	×	0	=	0
1	×	1	=	1
1	×	2	=	2
1	×	3	=	3
1	×	4	=	4
1	×	5	=	5
1	×	**6**	=	**6**
1	×	7	=	7
1	×	8	=	8
1	×	9	=	9
1	×	10	=	10

= **6**

Learn to Multiply by 2

Two (**2**) puppies were walked by each of the **3** puppy walkers. How many puppies were walked by puppy walkers?

$$2 \times 3$$

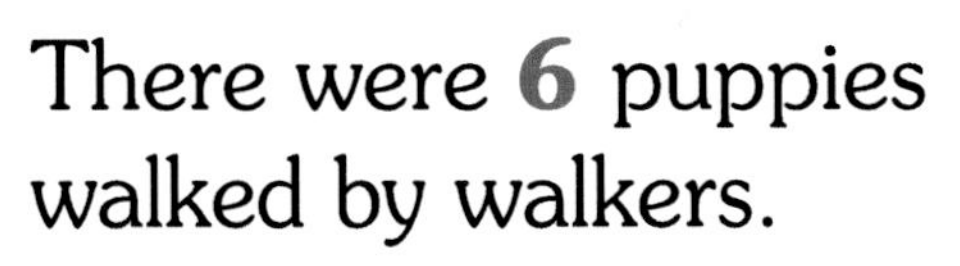

There were **6** puppies walked by walkers.

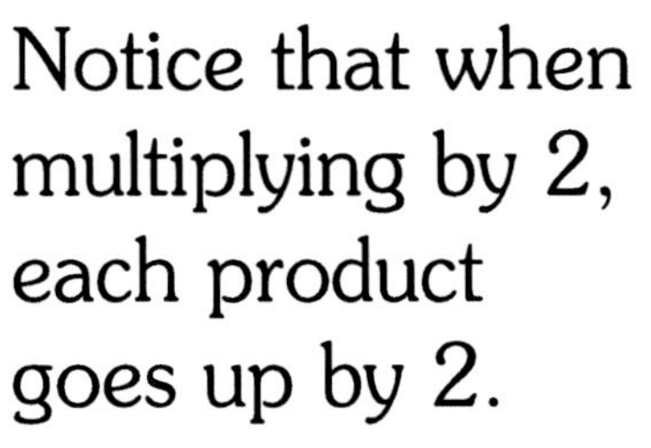

Notice that when multiplying by 2, each product goes up by 2.

Number Family

Multiplication and division are the reverse of each other and can help you learn the basic facts. For example:

$2 \times 3 = 6$ $\quad$ $6 \div 2 = 3$

$3 \times 2 = 6$ $\quad$ $6 \div 3 = 2$

This is called a number family.

$2 \times 0 = 0$
$2 \times 1 = 2$
$2 \times 2 = 4$
$\mathbf{2 \times 3 = 6}$
$2 \times 4 = 8$
$2 \times 5 = 10$
$2 \times 6 = 12$
$2 \times 7 = 14$
$2 \times 8 = 16$
$2 \times 9 = 18$
$2 \times 10 = 20$

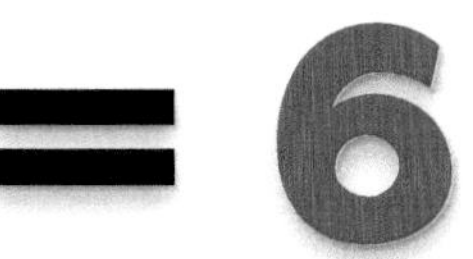

Learn to Multiply by 3

Three (**3**) children are each cuddling **4** kittens at once. How many kittens are there in all?

3  **4**

Repeated Addition

Multiplication can be thought of as repeated addition.

$3 \times 5 = 15$

$3 + 3 + 3 + 3 + 3 = 15$

Adding 3 five times equals 15.

3	×	0	=	0
3	×	1	=	3
3	×	2	=	6
3	×	3	=	9
3	×	**4**	=	**12**
3	×	5	=	15
3	×	6	=	18
3	×	7	=	21
3	×	8	=	24
3	×	9	=	27
3	×	10	=	30

There are **12** kittens in all to cuddle.

= 12

Learn to Multiply by 4

There are **4** paws on each of the **6** puppies. How many paws do the puppies have altogether?

4 × 6

Checking Your Work

Remember that multiplication and division are the reverse of each other. So, multiplication can be checked by dividing the product by either of the factors. If your answer is the other factor, then your multiplication is correct.

$$24 \div 4 = 6 \text{ AND } 24 \div 6 = 4$$

The puppies have **24** paws in all.

$4 \times 0 = 0$
$4 \times 1 = 4$
$4 \times 2 = 8$
$4 \times 3 = 12$
$4 \times 4 = 16$
$4 \times 5 = 20$
$\mathbf{4 \times 6 = 24}$
$4 \times 7 = 28$
$4 \times 8 = 32$
$4 \times 9 = 36$
$4 \times 10 = 40$

= 24

Learn to Multiply by 5

Five (**5**) feathers are floating around a room. Two (**2**) kittens are chasing each feather. How many kittens are chasing feathers?

5 × 2

Ten (**10**) kittens are chasing feathers.

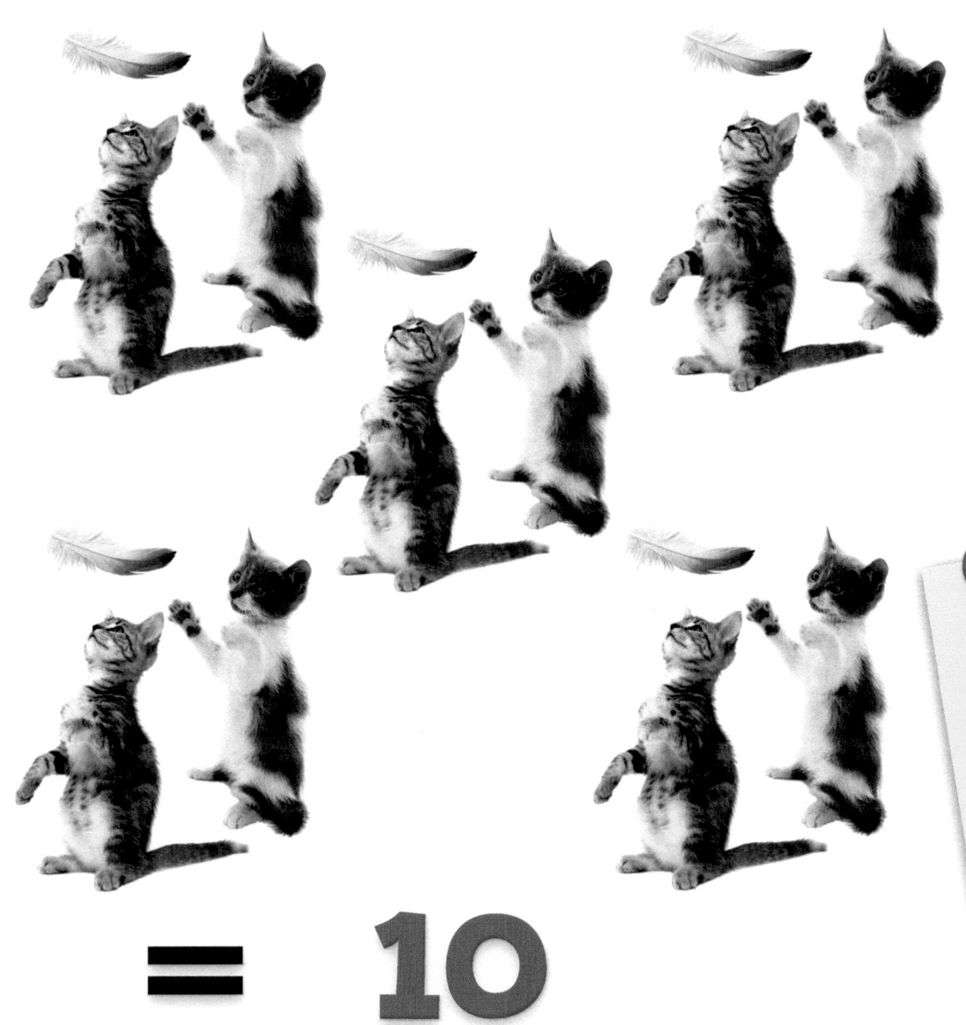

= **10**

5	×	0	=	0
5	×	1	=	5
5	**×**	**2**	**=**	**10**
5	×	3	=	15
5	×	4	=	20
5	×	5	=	25
5	×	6	=	30
5	×	7	=	35
5	×	8	=	40
5	×	9	=	45
5	×	10	=	50

Fantastic Fives!

Notice when you multiply by 5, it is like counting by 5s! Easy! The products are all numbers that end in 5 or 0.

Learn to Multiply by 6

Six (**6**) puppies each have **3** puppy treats. How many puppy treats do the puppies have in all?

6 × 3

The puppies have **18** treats altogether.

Try this with beads or counters of different colors. Group **6** of each color, then count up how many there are altogether.

= 18

$6 \times 0 = 0$
$6 \times 1 = 6$
$6 \times 2 = 12$
$6 \times 3 = 18$
$6 \times 4 = 24$
$6 \times 5 = 30$
$6 \times 6 = 36$
$6 \times 7 = 42$
$6 \times 8 = 48$
$6 \times 9 = 54$
$6 \times 10 = 60$

Associative Property

You can solve $2 \times 3 \times 3$ by first finding $2 \times 3 = 6$, then $6 \times 3 = 18$, or by finding $3 \times 3 = 9$, then $2 \times 9 = 18$. This is called the associative property of multiplication.

Learn to Multiply by 7

Seven (**7**) kittens each had **3** cat toys to play with. How many cat toys did they have between them all?

7 × 3

The kittens had **21** cat toys to play with.

$7 \times 0 = 0$
$7 \times 1 = 7$
$7 \times 2 = 14$
$\mathbf{7 \times 3 = 21}$
$7 \times 4 = 28$
$7 \times 5 = 35$
$7 \times 6 = 42$
$7 \times 7 = 49$
$7 \times 8 = 56$
$7 \times 9 = 63$
$7 \times 10 = 70$

= 21

Distributive Property

If you know $7 \times 2 = 14$ and $7 \times 3 = 21$, you can solve 7×5 by writing it as $7 \times (2 + 3) = (7 \times 2) + (7 \times 3) = 14 + 21 = 35$. This is called the distributive property of multiplication.

Learn to Multiply by 8

Eight (**8**) puppies each chewed on **2** shoes. How many shoes got chewed up?

8  **2**

The puppies chewed up **16** shoes.

= 16

8	×	0	=	0
8	×	1	=	8
8	×	**2**	=	**16**
8	×	3	=	24
8	×	4	=	32
8	×	5	=	40
8	×	6	=	48
8	×	7	=	56
8	×	8	=	64
8	×	9	=	72
8	×	10	=	80

Learn to Multiply by 9

There were **9** cushy cat beds with **4** kittens snuggled in each one. How many kittens were snuggled in the cat beds?

9 × 4

Nifty Nines

There is a very interesting pattern in the products of the 9s facts. Starting with 9 x 1, each digit of each product adds up to 9: 9 = 9, 1 + 8 = 9, 2 + 7 = 9, 3 + 6 = 9, and 4 + 5 = 9. This can help you remember your multiplication facts.

Thirty-six (**36**) kittens were snuggled in the beds.

9 × 0 = 0
9 × 1 = 9
9 × 2 = 18
9 × 3 = 27
9 × 4 = 36
9 × 5 = 45
9 × 6 = 54
9 × 7 = 63
9 × 8 = 72
9 × 9 = 81
9 × 10 = 90

Learn to Multiply by 10

There are **10** puppies playing in each of the **2** kiddie pools. How many puppies are playing in the pools in all?

$$10 \times 2$$

Twenty (**20**) puppies are playing in the pools in all.

= 20

10 ×	0 =	0
10 ×	1 =	10
10 ×	**2 =**	**20**
10 ×	3 =	30
10 ×	4 =	40
10 ×	5 =	50
10 ×	6 =	60
10 ×	7 =	70
10 ×	8 =	80
10 ×	9 =	90
10 ×	10 =	100

Multiplication Tip for Tens

Do you see a pattern in the products here? Every time you multiply a number by 10, you can just add a 0 and you will have the correct product. This works for higher numbers, too!

Review Multiplication Facts

This is a multiplication table. Do you know all the facts in this book? Can you see the patterns on this table? Take a look!

X	0	1	2	3	4
0	0	0	0	0	0
1	0	1	2	3	4
2	0	2	4	6	8
3	0	3	6	9	12
4	0	4	8	12	16
5	0	5	10	15	20
6	0	6	12	18	24
7	0	7	14	21	28
8	0	8	16	24	32
9	0	9	18	27	36
10	0	10	20	30	40

Some boxes are highlighted in color. The blue boxes are multiples of **10**, the green boxes are multiples of **5**, and the yellow boxes are multiples of **3**. Notice there are some overlaps. You can use number facts you know to help you figure out number facts you don't know.

5	6	7	8	9	10
0	0	0	0	0	0
5	6	7	8	9	10
10	12	14	16	18	20
15	18	21	24	27	30
20	24	28	32	36	40
25	30	35	40	45	50
30	36	42	48	54	60
35	42	49	56	63	70
40	48	56	64	72	80
45	54	63	72	81	90
50	60	70	80	90	100

Activities with Multiplication

Play a Multiplication Game

You need paper, pencil, and one die. Roll the die and draw a shape on your paper for each dot on the die (big enough to draw other shapes inside). Roll the die again and draw another shape for each dot on the die inside each of the other shapes. You can use any simple shapes, such as squares and circles, or triangles and stars. Simply count up all the smaller shapes inside the bigger shapes and write a number sentence. For example: **4** squares each have **3** circles inside them. That is **12** circles in all. Write: **4 × 3 = 12** and **3 × 4 = 12**. You can even add division facts to make a complete fact family! **12 ÷ 3 = 4** and **12 ÷ 4 = 3**. Put the papers together and create a book!

Make Flash Cards

Write a multiplication fact on an index card and put the product on the other side. Make as many flash cards as you want. Test yourself! How many did you get right?

Look for Patterns

Look at the multiplication table in this book for patterns. Notice the products that repeat. They may have different factors or just reversed factors. How many can you find? Look for patterns out in the world, too. You will begin to notice things in multiples, such as packages of markers or cookies. How many things are there in two packages? In three packages?

Learn More

Books

Brack, Amanda, and Sky Pony Press. ***Math for Minecrafters: Adventures in Multiplication and Division.*** New York, NY: Sky Pony Press, 2017.

Midthun, Joseph. ***Multiplication.*** Chicago, IL: World Book, Inc., 2016.

Thinking Kids. ***Brainy Book of Multiplication and Division.*** Greensboro, NC: Thinking Kids, 2015.

Williams, Zella, and Rebecca Wingard-Nelson. ***Word Problems Using Multiplication and Division.*** New York, NY: Enslow Publishing, 2017.

Websites

Doctor Genius, Multiplication
www.mathabc.com/math-3rd-grade/division
Solve multiplication problems online!

Math-Play.com, 3rd Grade Math Games
www.math-play.com/3rd-grade-math-games.html
Practice your multiplication skills with more games!

Math Playground
www.mathplayground.com/games2.html
Play fun games to learn multiplication!

Index

To Eliza and Wiley, who taught me who I am

Published in 2018 by Enslow Publishing, LLC.
101 W. 23rd Street, Suite 240, New York, NY 10011

Library of Congress Cataloging-in-Publication Data

Names: Baker, Linda R., author.
Title: Learning multiplication with puppies and kittens / Linda R. Baker.
Description: New York, NY : Enslow Publishing, 2018. | Series: Math fun with puppies and kittens | Audience: K to grade 3. | Includes bibliographical references and index.
Identifiers: LCCN 2017018121 | ISBN 9780766090910 (library bound) | ISBN 9780766090729 (pbk.) | ISBN 9780766090774 (6 pack)
Subjects: LCSH: Multiplication—Juvenile literature. | Mathematics—Juvenile literature.
Classification: LCC QA115 .B354 2018 | DDC 513.2/14—dc23
LC record available at https://lccn.loc.gov/2017018121

Printed in China

To Our Readers: We have done our best to make sure all websites in this book were active and appropriate when we went to press. However, the author and the publisher have no control over and assume no liability for the material available on those websites or on any websites they may link to. Any comments or suggestions can be sent by email to customerservice@enslow.com.

Photo credits: Cover, p. 1 (puppy, left) PCHT/Shutterstock.com; cover, p. 1 (puppy, center) cynoclub/Shutterstock.com; cover, pp. 1 (kitten), 2 (last 6 animals from the right), 18 (puppy), 19 (puppy) Ermolaev Alexander/Shutterstock.com; pp. 2 (kitten, left), 16 (grey and white kitten), 17 (grey and white kitten) Happy monkey/Shutterstock.com; p. 2 (fluffy puppy) Liliya Kulianionak/Shutterstock.com; p. 4 otojagodka/Thinkstock.com; p. 5 Tatiana Katsai/Shutterstock.com; p. 7 cynoclub/Thinkstock.com; pp. 8, 9 (cat tree) happymay/Shutterstock.com; pp. 8. 9 (kitten) Utekhina Anna/Shutterstock.com; p. 9 (puppy) Kalamurzing/Shutterstock.com; p. 10 (puppy) Gelpi/Shutterstock.com; pp. 10 (woman), 11 (all) ESB Professional/Shutterstock.com; pp. 12 (boy, left), 13 (boy, left) michaeljung/Shutterstock.com; pp. 12 (girl, center), 13 (girl, center) Di Studio/Shutterstock.com; pp. 12 (boy, right), 13 (boy, right) all_about_people/Shutterstock.com; pp. 12 (kittens), 13 (kittens) zakharov aleksey/Shutterstock.com; p. 13 (puppy) ARTSILENSE/Shutterstock.com; p. 14 (paw) oksana2010/Shutterstock.com; pp. 14 (puppy), 15 (puppy) OllyLo/Shutterstock.com; pp. 16 (feather), 17 (feather) kzww/Shutterstock.com; pp. 16 (striped kitten), 17 (striped kitten) Tony Campbell/Shutterstock.com; pp. 18 (treats), 19 (treats) Charles Brutlag/Shutterstock.com; pp. 20, 21 (all kittens) Tuzemka/Shutterstock.com; pp. 20 (toys), 21 (toys) Ivonne Wierink/Shutterstock.com; pp. 22 (puppy), 23 (puppy with sneakers) Africa Studio/Shutterstock.com; p. 22 (sneakers) sergign/Shutterstock.com; pp. 24 (bed), 25 (bed) Vitaly Titov/Shutterstock.com; pp. 24 (group of kittens), 25 (group of kittens) Tsekhmister/Shutterstock.com; p. 25 (kitten with paw up) Lubava/Shutterstock.com; pp. 26 (puppy), 27 (puppy) Dorottya Mathe/Shutterstock.com; pp. 26 (pool), 27 (pool) Kittibowornphatnon/Shutterstock.com; p. 27 (kitten) yevgeniy11/Shutterstock.com.